# YOUR KNOWLEDGE HAS VALUE

# The Mathematics of Ranked-Choice Single-Winner Voting Systems. Can Different Systems of Voting Affect the Results?

Maciej Nodzyński

**Bibliographic information published by the German National Library:**

The German National Library lists this publication in the National Bibliography; detailed bibliographic data are available on the Internet at http://dnb.dnb.de.

ISBN: 9783346473509
This book is also available as an ebook.

© GRIN Publishing GmbH
Nymphenburger Straße 86
80636 München

Print and binding: Books on Demand GmbH, Norderstedt, Germany
Printed on acid-free paper from responsible sources.

The present work has been carefully prepared. Nevertheless, authors and publishers do not incur liability for the correctness of information, notes, links and advice as well as any printing errors.

GRIN web shop: https://www.grin.com/document/1059528

# Mathematics Applications and Interpretations

## Standard Level

The Mathematics of ranked-choice single-winner voting systems.

## 1.Introduction

It is not an exaggeration to say that mathematics is omnipresent in the surrounding world. As far as the eye can see, there are plenty of signs of mathematical activity. Consequently, there is no surprise that this science can also be found in politics and, more specifically, it has a lot in common with various types of elections.When I am writing this essay, exactly one month has passed since the last presidential elections in my home country, Poland, which for many weeks engaged the attention of citizens and caused extreme emotions. The elections were marginally won by Andrzej Duda, who became the President of Poland for the second term. I started wondering whether it is possible to change the results of a single-winner election by applying a different voting system.

That is why the aim of this paper is to investigate to what extent different systems of voting can affect the results and the distribution of votes in single-winner elections. Moreover, during elections, there are always politicians who hold extreme beliefs and are either loved or hated by people and those who hold tempered beliefs but are tolerable by society. Consequently, the research also examines which of the two mentioned characteristics a candidate should have to increase the chances of winning by applying basic statistical measures such as mean or standard deviation. The subject is relevant because it allows us to get a mathematical insight into single-winner electoral systems and can show whether a particular system of voting is only a tool, or directly contributes to the results. Moreover, I am deeply passionate about politics and I am curious whether it is doable to manipulate the results of the election by applying a particular voting system. I have decided to analyze 4 different methods of voting that are or were used in the world to elect a President or other representatives for single-member posts: Supplementary Vote (SV), Instant-runoff voting (IRV), Bucklin Voting, and Coombs' method. It is worth adding that these systems are ranked-choice systems, that is voters rank their candidates from the most favorable one to the one that in their opinion is completely not suitable to accede to a particular office. Ranked-choice systems were chosen because this paper does not investigate only who wins the election but also how the situation of candidates in other positions changes.

## 2. Procedure

The investigation started by conducting an anonymous poll which had to gather data needed for further analysis. Voluntary sampling was used because of its simplicity and low time consumption. The participants had to be over 18 years old as in Poland this is a requirement to be legally allowed to vote. The majority of data was gathered using an anonymous, online poll where the responses were obtained from people aged from 18 to 56 years old. The rest of the data was obtained from older people using the same, but paper-based survey. It was done to avoid a non-response bias, at the same time improving the representativeness of the poll. Moreover, the reliability was increased by the fact that there was a similar representation of each age group so it was not possible that one age group could distort the results as all groups were of a similar size. Fifty four participants ranked the candidates according to their preferences. The 6 candidates were the most known politicians who applied for the position of the President of Poland in 2020: Andrzej Duda, Rafał Trzaskowski, Szymon Hołownia, Krzysztof Bosak, Władysław Kosiniak-Kamysz, and Robert Biedroń who were marked with letters A, B, C, D, E, and F respectively to make the research more clear and transparent. These candidates were chosen because the poll was conducted shortly after the presidential elections, so the majority of participants have already formed an impression about each of the mentioned candidates and therefore, they did not have any problems to rank them. The main plan for this investigation is to use the data collected in the poll to explain how particular systems of voting work and investigate how the results differ among these systems. The pictures below show an online ballot the participants had to fill out (left-most) and a sample ballot from the Instant run-off election[1] (right-most).

### 1. Change the order of the candidates according to your preferences (1-the best, 6-the worst).*

⌃ 1. Andrzej Duda

⌃ 2. Rafał Trzaskowski

⌃ 3. Szymon Hołownia

⌃ 4. Krzysztof Bosak

⌃ 5. Władysław Kosiniak-Kamysz

⌃ 6. Robert Biedroń

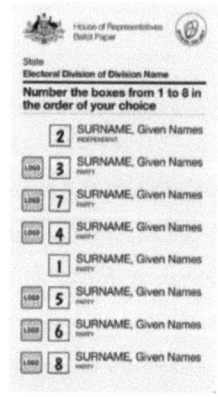

Picture 1: A picture showing an online ballot used to collect data.

Picture 2: A picture showing a sample ballot used in the ranked-choice elections in Australia.

The results of the survey are summarized in the tables below. The results of each candidate are presented in a separate table - only ballots where a particular candidate (marked as a letter) was the 1st choice, are included in respective tables. It was done to make the process of analysis easier and better-organized. As a result, 6 tables were created. At this stage, colors do not have any significance.

---

[1] Source: www.aec.gov.au

Candidate A

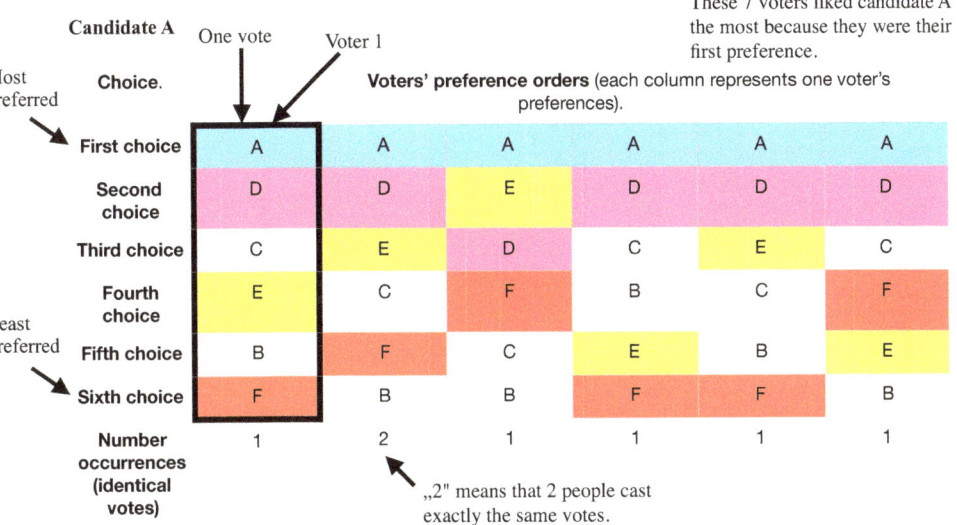

**Candidate A**  One vote  Voter 1

These 7 voters liked candidate A the most because they were their first preference.

Most preferred

Least preferred

| Choice. | Voters' preference orders (each column represents one voter's preferences). | | | | | |
|---|---|---|---|---|---|---|
| First choice | A | A | A | A | A | A |
| Second choice | D | D | E | D | D | D |
| Third choice | C | E | D | C | E | C |
| Fourth choice | E | C | F | B | C | F |
| Fifth choice | B | F | C | E | B | E |
| Sixth choice | F | B | B | F | F | B |
| Number occurrences (identical votes) | 1 | 2 | 1 | 1 | 1 | 1 |

"2" means that 2 people cast exactly the same votes.

Table 1: Table showing the distribution of votes when the voters' first choice was candidate A.

**Candidate B**

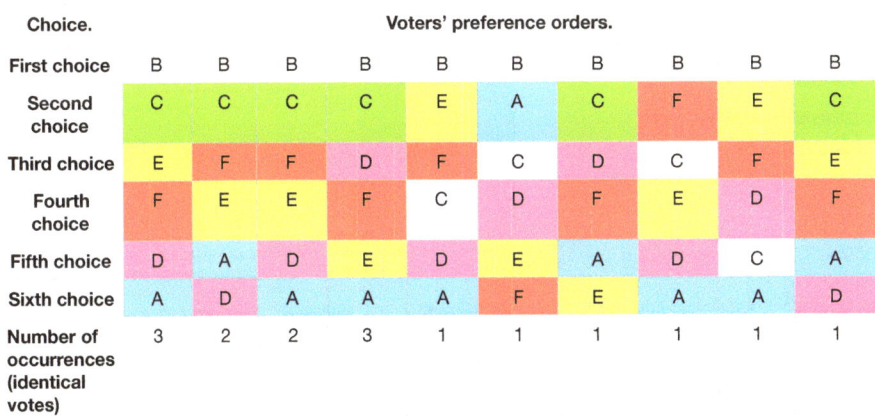

| Choice. | Voters' preference orders. | | | | | | | | | |
|---|---|---|---|---|---|---|---|---|---|---|
| First choice | B | B | B | B | B | B | B | B | B | B |
| Second choice | C | C | C | C | E | A | C | F | E | C |
| Third choice | E | F | F | D | F | C | D | C | F | E |
| Fourth choice | F | E | E | F | C | D | F | E | D | F |
| Fifth choice | D | A | D | E | D | E | A | D | C | A |
| Sixth choice | A | D | A | A | A | F | E | A | A | D |
| Number of occurrences (identical votes) | 3 | 2 | 2 | 3 | 1 | 1 | 1 | 1 | 1 | 1 |

Table 2: Table showing the distribution of votes when the voters' first choice was candidate B.

**Candidate C**

| Choice. | Voters' preference orders. | | | | | | | | | | | |
|---|---|---|---|---|---|---|---|---|---|---|---|---|
| First choice | C | C | C | C | C | C | C | C | C | C | C | C |
| Second choice | B | B | B | F | B | D | E | B | D | E | B | B |

| Third choice | E | F | F | B | E | E | D | E | B | B | D | D |
| --- | --- | --- | --- | --- | --- | --- | --- | --- | --- | --- | --- | --- |
| Fourth choice | F | E | A | A | F | F | B | D | F | F | E | E |
| Fifth choice | A | D | E | D | D | A | A | F | A | D | F | A |
| Sixth choice | D | A | D | E | A | B | F | A | E | A | A | F |
| Number of occurren-ces (identical votes) | 1 | 1 | 1 | 1 | 2 | 1 | 1 | 1 | 1 | 1 | 1 | 1 |

Table 3: Table showing the distribution of votes when the voters' first choice was candidate C.

**Candidate D**

| Choice | Voters' preference orders. | | | | | |
| --- | --- | --- | --- | --- | --- | --- |
| First choice | D | D | D | D | D | D |
| Second choice | C | E | B | A | C | A |
| Third choice | E | B | C | C | E | B |
| Fourth choice | F | F | E | E | B | E |
| Fifth choice | B | C | F | F | F | C |
| Sixth choice | A | A | A | B | A | F |
| Number of occurrences (identical votes) | 1 | 1 | 1 | 1 | 1 | 1 |

Table 4: Table showing the distribution of votes when the voters' first choice was candidate D.

**Candidate E**

| Choice. | Voters' preference orders. | | |
| --- | --- | --- | --- |
| First choice | E | E | E |
| Second choice | F | F | C |
| Third choice | C | C | D |
| Fourth choice | B | D | B |
| Fifth choice | D | B | A |
| Sixth choice | A | A | F |
| Number of occurrences (identical votes) | 1 | 1 | 1 |

Table 5: Table showing the distribution of votes when the voters' first choice was candidate E.

**Candidate F**

| Choice. | Voters' preference orders. | | | | | | | |
|---|---|---|---|---|---|---|---|---|
| **First choice** | F | F | F | F | F | F | F | F |
| **Second choice** | B | C | C | E | B | B | D | C |
| **Third chice** | C | B | B | B | E | C | B | E |
| **Fourth choice** | E | E | E | C | D | E | C | B |
| **Fifth choice** | D | A | D | A | C | D | E | D |
| **Sixth choice** | A | D | A | D | A | A | A | A |
| **Number of occurrences (identical votes)** | 2 | 1 | 1 | 1 | 1 | 1 | 1 | 1 |

Table 6: Table showing the distribution of votes when the voters' first choice was candidate F.

### 3. Analysis of voting systems

- **First past the post (Plurality voting)**

Although this system does not belong to ranked voting systems and is not mathematically complicated, it is an excellent introduction to the systems described in the later part of this paper. The first past the post (FPTP) system, defined also as a plurality voting, functions as a way to elect a president in over 20 countries, for example in Honduras, Iceland, or Singapore. FPTP is a simple and logical method based on two main rules by which every citizen casts only one vote and the candidate who obtained the most votes wins.

**Investigation**

Firstly, in the tables from section 2 the candidate with most first-choice votes can be found. This candidate is the winner of these elections (Candidate B-16 votes). The rest of the candidates have been ranked according to this criterion.

**Final results**

| Place | Candidate | Number of first-choice votes (out of 54) |
|---|---|---|
| 1st | B | 16 |
| 2nd | C | 13 |
| 3rd | F | 9 |
| 4th | A | 7 |

| 5th | D | 6 |
| 6th | E | 3 |

Table 1: Table showing how the results of the election would look like if the FPTP method had been used. Although FPTP is a simple and time-saving method it has serious disadvantages. The main drawback is that the candidate can win presidential elections by having the plurality (the highest number of votes), not the majority (over 50%), of votes. The best example of it is the presidential election won by Fidel Ramos with only 24 % of the popular vote in the Philippines in 1992. It was enough that he had the plurality, although the rest of entitled to vote citizens voted for the other 6 candidates. The next issue concerns the point of view of extremists who are usually the minorities ignored by the FPTP method. Therefore, FPTP is suitable for elections where only 2 candidates compete with each other.

- **Supplementary vote**

The first preferential voting system called Supplementary Vote (SV) is applied during the London mayoral election. The voters are given a ballot with two columns. In the first column, they must mark their favorite candidate whereas in the second one they can choose their second preference. The election is won by a candidate with over 50% of votes (in this case, the majority is 28 or more votes). If none of the candidates receives the majority, the top two candidates compete with each other in the second run. The rest of them are eliminated, however, the second choice is still important. If the second preferred candidate is one of the top two candidates, then this vote can be regarded as an extra one for a particular candidate.

**Investigation**

Consider the first preferences to check whether a certain candidate obtained the majority of the first-choice votes. Table 1 from section 3 shows that there is no winner in the first round, but there are two top candidates: candidate B (16 votes) and C (13 votes). The remaining candidates are eliminated, however, the second preferences of these candidates' voters are taken into account. The second-choice votes for the top two candidates are colored in green in tables from section 2.

Candidate B: 16 first-place votes +12 second-place votes = 28 votes, the majority

Candidate C: 13 first-place votes+18 second-place votes = 31 votes, the **winner** with a greater majority of votes

**Important note:** In this part of the paper the reader can come across on words „the majority" and „the greater majority" as mentioned above which may be confusing and illogical at first sight. Therefore, I would like to rectify that „the majority" always means 28 votes or more, no matter how many next-preference votes are added. The majority is equal to 28 and is constant because there were 54 voters. If there are at least 2

7

candidates who obtained 28 votes or more, the elections are won by the candidate with the „greater majority"
of votes, that is the candidate who obtained the most votes. For example, in the Supplementary Vote system
described above, there are top two candidates and each of them has the majority, that is at least 28 votes.
However, candidate C has more votes than candidate B and therefore, has the greater majority of votes. This
rule applies to every voting system described in this paper.

**Final results**

| Place | Candidate | Number of first-choice votes (out of 54) |
|---|---|---|
| 1st | C | 31 |
| 2nd | B | 28 |
| 3rd | F | 9 |
| 4th | A | 7 |
| 5th | D | 6 |
| 6th | E | 3 |

Table 2:  Table showing how the results of the election would look like if the Supplementary vote method
had been used.

The asset of this system is that it is simple to understand. Moreover, the voter has more power since both
their first and second choices might count whereas the majority criterion prevents from obtaining the
confusing results, as in the FPTP method. On the other hand, SV may involve tactical voting because the
second preferences may also be important. Furthermore, there are a lot of „wasted votes" as the first choices
of candidates who are eliminated in the first round no longer count, only second choices may. Additionally, if
a voter is not sure whether there will be a winner after the first round, they may try to second-guess which
two candidates will be in the second round and allocate their votes in such a way to make their prediction
true. In such situation there is a possibility that a voter will inadvertently beat their favorite candidate. Table
2 indicates that in this method a candidate should have a similar number of first and second-choice votes to
win rather than trying to get as many first-choice votes as possible.

- **Instant-runoff voting (IRV or Plurality with elimination)**

The following system is used to elect the President of Ireland and forms an extended version of FPTP.
Firstly, the voters have to rank the candidates according to their preferences using numbers, starting with
using "1" next to their favorite candidate on the ballot. If over half of the voters ranked the same candidate in
the first place, then this candidate wins. If not, the candidate who obtained the fewest first-place votes is

eliminated. However, the votes of those who voted for this candidate are still valid and move to the one who was their next preference. This procedure is repeated until one of the candidates has the majority and wins the election. If two or more candidates obtain the majority, the candidate with the greater majority wins. Here is a simple explanation of how it works in practice.

| First choice | Candidate B |
|---|---|
| Second choice | Candidate F |
| Third choice | ~~Candidate D~~ Transfer of 2 votes |
| Fourth choice | Candidate C |
| Number of votes | 2 |

Let's assume that candidate D obtained the fewest first-place votes. They are eliminated and the votes for they move to the next, undermentioned preference.

**This candidate is eliminated from the elections.**

Table 3: Table explaining the mechanism of the IRV method.

It is worth adding that if a candidate takes the last place, votes for them are lost since there is no preference they could be added to.

**Investigation**

Determine the candidate with the fewest number of first-place votes. Table 1 indicates that this is candidate E (3 first-place votes), hence, they are eliminated. Their votes are highlighted in yellow in tables in section 2.

Candidate A: 7 first-place votes

Candidate B: 16 first-place votes

Candidate C: 13 first-place votes

Candidate D: 6 first-place votes

Candidate E: 3 first-place votes, **1st eliminated**

Candidate F: 9 first-place votes

Move the votes from Candidate E to the candidate that was the next preference.

Candidate A: 7+8 = 15 votes , **2nd eliminated**

Candidate B: 16+7 = 23 votes

Candidate C: 13+5 = 18 votes

Candidate D: 6+13 = 19 votes

Candidate E: 3, **1st eliminated**

Candidate F: 9+18 = 27, a half of the number of votes, but not the majority.

No one obtained the majority. Eliminate the next candidate with the fewest number of first-place votes (candidate A-15 votes). The votes for them are highlighted in blue in tables in section 2. Move Candidate A's votes to the candidate who was the next preference (it should be bearing in mind that votes for Candidate E have already been used).

Candidate B: 23+2 = 25

Candidate C: 18+2 = 20

Candidate D: 19+14 = 33, The **winner** with a greater majority of votes

Candidate F: 27+3 = 30, the majority

**Final results**

| Place | Candidate | Number of votes |
|-------|-----------|-----------------|
| 1st | D | 33 |
| 2nd | F | 30 |
| 3rd | B | 25 |
| 4th | C | 20 |
| 5th | A | 15 |
| 6th | E | 3 |

Table 4:  Table showing how the results of the election would look like if the IRV method had been used. On one hand, IRV is a good method because the winner must obtain the majority and, on account of the complexity of the system, it is difficult to involve tactical voting. The voters have much more power since many of their preferences may count and decide who wins the elections. On the other hand, it is unclear whether IRV favors candidates from small or big parties since table 4 shows that candidates with fewer first-choice votes were placed at the beginning (D, F) and the end (A, E) of the table whereas the candidates with a lot of first-choice votes were placed in the middle (B, C). The next issue is that voters do not have to rank all candidates, especially the ones that they are not fond of. If enough voters do not rank the candidates on their lower preferences, none of the candidates may obtain the majority. That is why Australia requires the voters to rank every candidate, however, some claim that it breaches the voting rights.

- **Bucklin voting (Grand Junction system)**

This system is similar to IRV, however, there are some differences. At the very beginning, voters rank the candidates according to their preference. Afterwards, the first-place votes are counted and if a certain candidate obtains the majority, they win. If not, the second choices are added to the first ones and again the votes are counted. The successive choices are added until the winner obtains the majority with the most votes

10

or there are no more choices to be added. The method was commonly used in American cities in the first half of 20th century (eg. Portland, Denver).

## Investigation

Determine the candidate who obtained the majority of first-place votes using tables from section 2. If none of the candidates obtained the majority, add the second-choice votes to the first-choice votes using the tables from section 2.

Candidate A: 7 first-place votes + 3 second place votes = 10 votes

Candidate B: 16 first-place votes + 13 second place votes = 29 votes, the majority

Candidate C: 13 first-place votes + 18-second place votes = 31 votes, the **winner** with a greater majority of votes

Candidate D: 6 first-place votes + 9 second-place votes = 15 votes

Candidate E: 3 first-place votes + 7 second-place votes = 10 votes

Candidate F: 9 first-place votes + 4 second-place votes = 13 votes

| Place | Candidate | Number of votes |
|-------|-----------|-----------------|
| 1st | C | 31 |
| 2nd | B | 29 |
| 3rd | D | 15 |
| 4th | F | 13 |
| 5th | A,E | 10 |

Table 5: Table showing how the results of the election would look like if the Bucklin method had been used. The first advantage of this system is that it uses the majority criterion and that every vote and preference may decide about the election results. However, in some cases, this method seems to be unfair. For example, if in the first round a certain candidate obtains the majority, they win the elections and other preferences are simply disregarded. Then, Bucklin voting method is very similar to the FPTP method. The only difference is that the Buckling method involves the majority criterion. The second example is when the votes were cast in such a way that the last choices are added. In my opinion, it is not reasonable that the least preferred candidate can benefit from being the last and win the election because of the last-place votes. To conclude, this system is unpredictable when it comes to the power that voters have. To win the elections, the best strategy is to have as many first and second-choice votes as possible because later the system may be very unpredictable.

- **Coombs' method**

The following system works in the same way as IRV, but in the reverse direction. Instead of eliminating candidates with the fewest first-place votes, only those with the most last-place votes are eliminated. The votes of the eliminated candidates move to the next preferences, as in the IRV method. The winner is the candidate who obtained the greatest majority of votes. The method has not been used in any official elections, nevertheless, it is worth investigating in which order the candidates will be placed and whether it will be the same order as in IRV.

**Investigation**

Determine the candidate with the most last-place votes using the tables from section two (candidate A-30 last-place votes) and eliminate them.

Candidate A: 30 last-place votes, **1st eliminated**

Candidate B: 6 last-place votes

Candidate C: 0 last-place votes

Candidate D: 7 last-place votes

Candidate E: 3 last-place votes,

Candidate F: 8 last-place votes

Move the votes for candidate A to the votes to the candidate that was the next preference. Candidate A's votes are highlighted in blue in tables in section 2.

Candidate B: $16+2 = 18$ votes

Candidate C: $13+2 = 15$ votes

Candidate D: $6+11 = 17$ votes

Candidate E: $3+2 = 5$ votes

Candidate F: $9+3 = 12$ votes

Because there is **no majority**, the procedure should be repeated (it is worth adding that votes for candidate A have been already used). Determine the candidate with the most last-place votes using tables from section 2 (candidate D-24 last-place votes) and eliminate them.

Candidate B: 8 last-place votes

Candidate C: 3 last-place votes

Candidate D: 24 last-place votes, **2nd eliminated**

Candidate E: 7 last-place votes

Candidate F: 12 last-place votes

Move Candidate D's votes to the next preferred candidate. Candidate D's votes are highlighted in pink.

Candidate B: 18+6 = 24 votes

Candidate C: 15+7 = 22 votes

Candidate E: 5+9 = 14 votes

Candidate F: 12+4 = 16 votes

Again, since there is **no majority**, the procedure is repeated (bearing in mind that votes for candidate A and D have already been used). Determine the candidate with the most last-place votes using tables from section 2 (candidate F-21 votes) and eliminate them.

Candidate B: 10 last-place votes

Candidate C: 4 last-place votes

Candidate E: 19 last-place votes

Candidate F: 21 last-place votes, **3rd eliminated**

Move Candidate F's votes to the next preferred candidate. Votes for Candidate F are highlighted in red.

Candidate B: 24+9 = 33 votes, The **winner** that has a greater majority of votes

Candidate C: 22+10 = 32 votes, the majority

Candidate E: 14+10 = 24 votes

**Final results**

| Place | Candidate | Number of votes |
|---|---|---|
| 1st | B | 33 |
| 2nd | C | 32 |
| 3rd | E | 24 |
| 4th | F | 16 |
| 5th | D | 17 |
| 6th | A | 7 |

Table 6 : Table showing how the results of the election would look like if the Coombs' method had been used.

The advantage of this system is that it uses the majority criterion and that it firstly eliminates the candidate with the larger number of last-place votes. It is reasonable since the least liked candidate is eliminated at the very beginning so there is a very small likelihood that the voters may be disappointed with election results. Moreover, it should be noticed that candidate E, who was always on the last place in every system, moved to the third position which may indicate that the Coombs' method favors neutral candidates, that is the ones who have many third-choice and fourth-choice (middle) votes. The other interesting observation is that candidate D has more votes than candidate F but is in the lower position in the table. It is because candidate D had more last-place votes than candidate F and therefore, he was eliminated first. Besides that, it should be mentioned that the system is vulnerable to tactical voting, for example to teaming which is when a candidate enters the election not to win, but to change the winner.

**4. Is it better to be a stable candidate and have a similar number of votes on several places or to be an unstable one and have a different number of votes on all places?**

Should a candidate be stable or unstable to win the elections when the ranked system of voting is applied? It is an interesting question and the answer can be obtained with the help of the mean and standard deviation. The mean shows on which place a particular candidate was usually located (therefore, the lower value of the mean, the bigger chances to win) whereas standard deviation represents the stability of a candidate, that is how data is clustered or distributed around the mean. I believe that the elections are fair when the winner has both the least value of the mean and standard deviation and this section will examine also the fairness of different systems of voting according to my criteria. The following formula for the mean will be used:

$$\mu = \frac{\sum x}{N}$$, where $\sum$ means the sum of x data values and $N$ is the number of data values. As for the

standard deviation: $\sigma = \sqrt{\dfrac{\sum |x - \mu|^2}{N}}$, where $N$ is the number of data values, $\mu$ is the mean of data

values and $x$ is the value in the data set. To calculate it, the first step is to obtain the mean of data values. Afterwards, we take the first value from the data set and subtract from it the value of the mean and square the difference. This procedure should be repeated with every data value. The next step is to add the squares and divide the sum by the number of data values. The last thing to do is to calculate the square root of the quotient and the result is standard deviation. However, which values should be used to calculate the standard deviation? I decided to construct one cumulative frequency graph for all candidates where the x-axis shows the possible places a candidate can take in the elections and the y-axis shows the frequency a candidate took a particular place in the elections. Each product from multiplying the values from the x-axis with the values from the y-axis is a data value used to calculate the standard deviation.

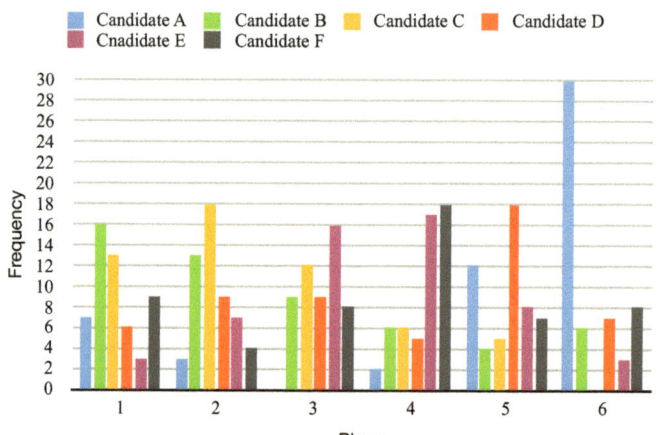

Graph 1: Graph showing the frequency voters placed candidates on particular places.

Here is how it works in practice for candidate A.

Firstly, I multiplied the values from the x-axis by the values from the y-axis.

$1 \times 7 = 7$

$2 \times 3 = 6$

$3 \times 0 = 0$

$4 \times 2 = 8$

$5 \times 12 = 60$

$6 \times 30 = 180$

Secondly, I calculated the mean.

$$\mu = \frac{7 + 6 + 0 + 8 + 60 + 180}{54} = 4.83$$

Thirdly, I calculate the distance of each data value from the mean and squared the difference.

$(4.83 - 7)^2 = 4.71$

$\left(4.83 - 6\right)^2 = 1.37$

$(4.83 - 0)^2 = 23.33$

$(4.83 - 8)^2 = 10.05$

$\left(4.83 - 60\right)^2 = 3043.73$

$(4.83 - 180)^2 = 30684.53$

Afterwards, I added these results and divided by 54, the number of data values.

$$\frac{4.71 + 1.37 + 23.33 + 10.05 + 3043.73 + 30684.53}{54} = 625.33$$

The last step is to take a square root from this value.

$$\sqrt{625.33} = 25.01$$

Therefore, the standard deviation is equal to **25.01**

I repeated this procedure for the rest of the candidates and the results are as follows:

Candidate B: $\mu = 2.76, \sigma = 7.65$

Candidate C: $\mu = 2.48, \sigma = 7.85$

Candidate D: $\mu = 3.76, \sigma = 13.55$

Candidate E: $\mu = 3.54, \sigma = 12.00$

Candidate F: $\mu = 3.63, \sigma = 12.24$

The results show that is it better to be a stable candidate since candidates B and C who have the lowest values of standard deviation were in the first 2 places in almost every system and candidate A with the higher value of standard deviation was always in the lower part of the table. The exception is the IRV method where candidate D, with a high value of standard deviation, managed to win the elections whereas candidates B and C were in the middle of the table. IRV seems to favor the candidates with other characteristics. The next argument for being stable is that 8 times at least one of the three candidates with the lowest value of standard deviation (B, C, E) were on the first 3 places while the rest of the candidates with higher values of standard deviation (A, D, F) were on the first 3 places only 5 times. Nevertheless, these are only my observations. To make my conclusions more reliable I decided to check whether there is a correlation between the value of standard deviation and the frequency of being on the first 3 places (podium) in ranked-choice voting systems, where the chances of winning are bigger. The Spearman's rank correlation coefficient ($r_s$) was used to check the correlation since the relationship between the ranks, not continuous data, had to be considered. Having used this technology, it turned out that the value of $r_s$ was equal to -0.81 which means that there was a considerable negative correlation. This indicates that the lower value of standard deviation, and therefore, larger stability, the larger the chance of taking one of the first places in election results.

As for the value of the mean, in the vast majority of candidates, the value of standard deviation was directly proportional to the value of the mean. It indicates some kind of correlation between these measures. To prove it, I decided to use the Pearson's product-moment correlation coefficient ($r_p$) to check the correlation between the value of standard deviation and the value of the mean. The value of $r_p$ was equal to 0.96 (strong positive correlation) which means that the lower value of the mean, the lower the standard deviation. It suggests that candidates who took first places have at the same time more clustered data (votes). That is why, to increase the chances of winning the election when ranked systems of voting are applied, a candidate should be first of all stable.

When it comes to the fairness of the results, in my opinion, the most fair results are obtained when the elections are won by the candidate who has both the least value of the mean and standard deviation or meets at least one of these requirements. When looking at the above calculations, it turns out that there are 2 most fair orders:

- According to the values of the standard deviation: BCEFDA, where candidate B is on the first position.
- According to the values of the mean: CBEFDA, where candidate C is on the first position.

If we look at the results of each ranked-choice voting system in section 3, it can be seen that the most fair system of voting is Coombs' method with the order of candidates BCEFDA (perfect order), where candidate B is in the first position. As for the most unfair system, it is the IRV method with the order of candidates DFBCAE, where candidate D is in the first position.

## 5. Investigating how the votes are distributed in different systems of voting.

Another interesting concept to explore is to show how data is distributed in every system of voting. I decided to use whisker and box diagrams since they enable us to see whether some systems tend to increase the difference between candidates or, on the contrary, decrease the difference. Moreover, they present the median- middle value, the interquartile range which shows how wide the scope of data is, and the extreme values which also show the variance of the data. The median is calculated using $\frac{n+1}{2}$ formula, where $n$ is the number of data values. After adding 1, the numerator is divided by 2 to find the median. If there is no value in the middle, a mean is taken from two middle values to obtain the median. As for the interquartile range (IQR), it allows us to see where the bulk of the data values lies. To calculate it, firstly a median should be obtained by dividing the data set in a half. Secondly, the median from each half is then taken. The middle value of the lower part is called the lower quartile (Q1) and the middle value of the upper part is called the upper quartile (Q3). When Q1 is subtracted from Q3, the IQR is obtained.

- **Supplementary vote**

- The minimum value: 3

- The maximum value: 31

- Range: $31 - 3 = 28$

- The median: 8

- The lower quartile: 4.5

- The upper quartile: 29.5

- The interquartile range (IQR): $29.5 - 4.5 = 25$

- The diagram:

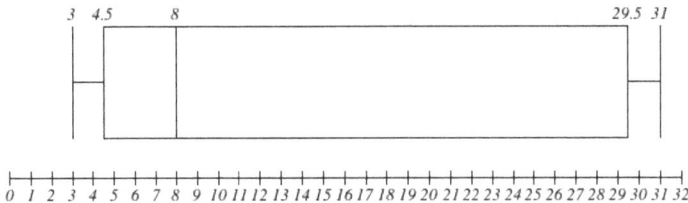

Diagram 1: A diagram showing the distribution of votes in Supplementary vote method.

- **Instant-runoff voting (Plurality with elimination)**

- The minimum value: 3

- The maximum value: 33

- Range: $33 - 3 = 30$

- The median: 22.5

- The lower quartile: 9

- The upper quartile: 31.5

- The interquartile range (IQR): $31.5 - 9 = 22.5$

- The diagram:

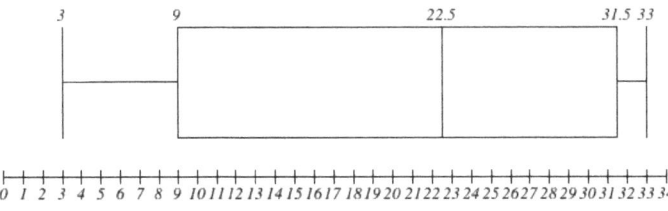

Diagram 2: A diagram showing the distribution of votes in Instant-runoff voting method.

18

- **Bucklin voting (Grand Junction system)**

- The minimum value: 10

- The maximum value: 31

- Range: $31 - 10 = 21$

- The median: 14

- The lower quartile: 10

- The upper quartile: 30

- The interquartile range (IQR): $30 - 10 = 20$

- The diagram:

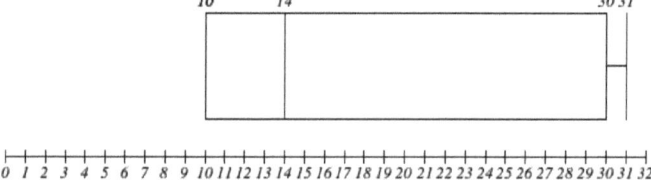

Diagram 3: A diagram showing the distribution of votes in Buckling voting method.

- **Coombs' method**

- The minimum value: 7

- The maximum value: 33

- Range: $33 - 7 = 26$

- The median: 20.5

- The lower quartile: 11.5

- The upper quartile: 32.5

- The interquartile range (IQR): $32.5 - 11.5 = 21$

- The diagram:

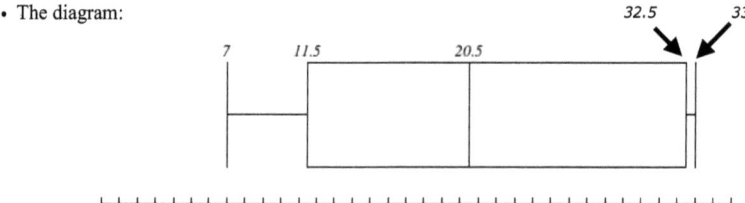

Diagram 4: A diagram showing the distribution of votes in Coombs' method.

19

The analysis of the diagrams shows several interesting aspects. When it comes to the Bucklin voting, this method had the lowest interquartile range (20) which means that the bulk of data was most clustered in this method. Moreover, Bucklin voting is also characterized by the smallest spread of data in general, as the range was also the lowest (21). It is also worth mentioning that the interquartile range was almost as long as the general range which means that there are no extreme values that could distort the results, as in the other methods. Regarding SV, in this method, the bulk of data was the most widely distributed since the interquartile range was the highest (25). SV had also the lowest median (8) whereas the IRV system had the highest median (22.5). Concerning the Coombs' method, the only thing worth mentioning is that the graph shows almost normal distribution, however, it does not bring anything relevant to this paper.

## 6. Evaluation

- **Conclusion**

The research showed that applying different methods of voting does influence the results of single-winner elections. In my investigation, at every turn, I obtained a different order of candidates. The winners turned out to be candidates B, C, and D. The case of candidate D is extremely interesting because according to 3 systems he should be in the penultimate place whereas the IRV method announced him as a winner of the election. One more surprising fact is that candidate E with only 3 first-place votes took the third position when Coombs's method was used. It proves that various systems of voting, especially the rank-choice ones, significantly influence the final results of the elections that can be won only by one candidate. Furthermore, the paper proposes an answer referring to the characteristics a candidate should have to win the election, where ranked-choice methods of voting are applied by using mean and standard deviation. Namely, my research revealed that a candidate should be first of all stable to win the elections. Moreover, although first-choice votes do have a significant impact, it seems that the most suitable strategy for the candidate in the case of ranked-choice systems is not to try to gain as many first-place votes as possible. It is crucial to be acceptable by citizens and attempt to be placed by them on the second or even third position on the ballot. This research examined also the fairness of voting systems and how every system distributes the votes, which is astonishing since the calculations and diagrams used in this paper showed these systems from the inside. One more thing worth noting is that a system of voting is not always a tool that only counts the votes but also a tool that can be used to manipulate the election results and favor a candidate of specific characteristics. To conclude, the mathematics of voting has truly enhanced my interest in this area and I was surprised how various results the systems of voting can bring.

- **Strengths, limitations, and suggestions for improvement.**

The biggest strength of this paper is that the subject is highly practical and applicable to real-life. Even though the research does not include many sophisticated equations or formulas, the investigation got insight into the mathematics of voting from the logical perspective, which is not less significant than the most equations-rich mathematics. Moreover, my investigation extends the prior knowledge about how voting systems work and how their mechanisms can affect the results of single-winner elections. Another advantage of this paper is that the structure and the content is transparent and understandable. The voting systems are explained appropriately in a written and graphical way what makes the paper easy to follow. The last asset is that the research gave truly thought-provoking and interesting results. Every system of voting determined election results.

On the other hand, there is room for improvement on several aspects. Firstly, I think I should have collected votes in such a way that each candidate would have an equal number of first-place votes because these votes always give a larger or smaller advantage and the research showed it. A good example is candidates B and C who took the first two places almost in every ranking. With respect to the sample, it was relatively small what could distort the results since voting systems usually proceed with a huge number of votes during elections. Voting systems may work differently when the sample is bigger. However, if the research was to be replicated I would recommend collecting fewer votes since there has been already too much data to analyze and too many relationships to research. Besides that, the sample improved the reliability of the study because the participants were not informed about the investigated systems of voting when they filled out the survey. It was done to prevent using any voting strategies which could have had an impact on the final results.

To conclude, if I were to replicate this investigation, I would have gathered data from about 30 participants and I would have made sure that every candidate obtained an equal number of first-place votes to make the results more objective. The potential further directions of this paper refer to the investigation on a larger scale and researching how the methods described in this paper as well as other systems of voting would process the collected data. Moreover, several systems I wrote about have modified versions that could be investigated as well. Additionally, a challenging extension for this research would be to investigate how different systems of voting would affect the results of multi-winner elections, for example to parliament or their equivalents in other countries or organizations.

## 7. Appendices

**Appendix 1:** Calculating the value of the Spearman's rank correlation coefficient ($r_s$) using technology.

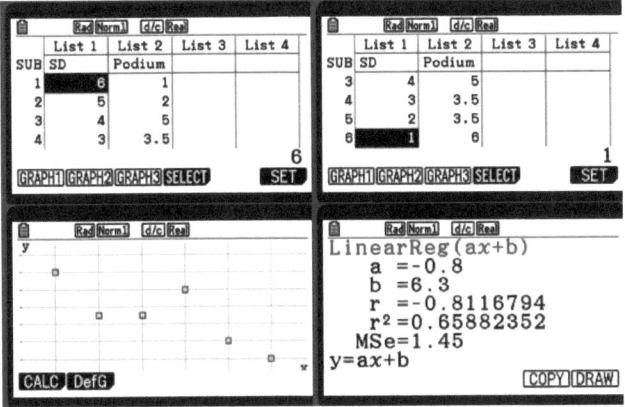

**Appendix 2:** Calculating the value of the Pearson's product-moment correlation coefficient ($r_p$) using technology.

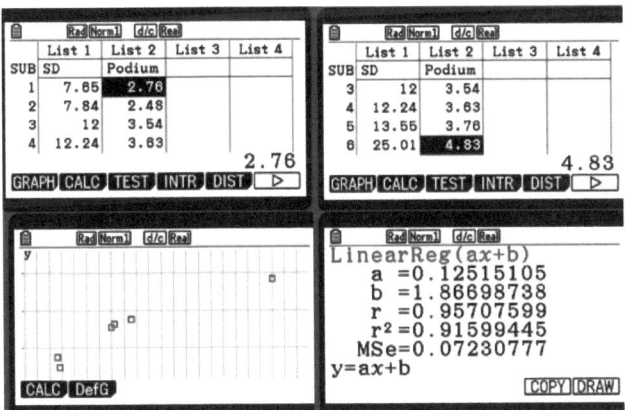

## 8. Bibliography

- Felt, A. and Natzke, C., 2016. *The Mathematics of Voting*. [video] Available at: <https://www.youtube.com/watch?v=AR2L7PR8Kf8> [Accessed 11 February 2021].

- Carneades.org, 2016. *Bucklin Voting (Voting System)*. [video] Available at: <https://www.youtube.com/watch?v=CkIYZsJAvNQ> [Accessed 11 February 2021].

- Think Mathematically, 2017. *Coombs Method of Voting*. [video] Available at: <https://www.youtube.com/watch?v=MSUltfMwnvI> [Accessed 11 February 2021].

- Electoral-reform.org.uk. 2017. *Voting Systems*. [online] Available at: <https://www.electoral-reform.org.uk/voting-systems/> [Accessed 11 February 2021].

- Aceproject.org. n.d. *Electoral Systems Index —*. [online] Available at: <https://aceproject.org/main/english/es/index.html> [Accessed 11 February 2021].

- Australian Electoral Commission. 2019. *Voting in the House of Representatives*. [online] Available at: <https://www.aec.gov.au/voting/how_to_vote/voting_hor.htm> [Accessed 12 February 2021].

- Hease, M., Humphries, M., Sangwin, C. and Vo, N., 2019. *Mathematics*. 1st ed. Marleston: Hease Mathematics.

- Hease, M., Humphries, M., Sangwin, C. and Vo, N., 2019. *Mathematics*. 1st ed. Marleston: Hease Mathematics.

- Imathas.com. n.d. *Boxplot Grapher*. [online] Available at: <https://www.imathas.com/stattools/boxplot.html> [Accessed 14 February 2021].